MW00565817

I Know
Big and Small

Reading consultant: Susan Nations, M.Ed.,
author, literacy coach, and consultant in literacy education

Photographs by Gregg Andersen

Printed in China

ISBN 13: 978-0-15-360211-5
ISBN 10: 0-15-360211-2

11 12 13 0940 16 15 14 13
4500409957

Harcourt
SCHOOL PUBLISHERS

Big tree

Small flower

Big elephant

Small mouse

Big dog

Small dog

Big shoes

Small shoes

Big truck

Small car

Glossary

big

small